Janete Ferreira da Silva

Socio-environmental Diagnosis of Sanitation in Campo Formoso-BA

Janete Ferreira da Silva

Socio-environmental Diagnosis of Sanitation in Campo Formoso-BA

Lack of environmental sanitation leads to the emergence of water-borne diseases and environmental degradation

ScienciaScripts

Imprint
Any brand names and product names mentioned in this book are subject to trademark, brand or patent protection and are trademarks or registered trademarks of their respective holders. The use of brand names, product names, common names, trade names, product descriptions etc. even without a particular marking in this work is in no way to be construed to mean that such names may be regarded as unrestricted in respect of trademark and brand protection legislation and could thus be used by anyone.

Cover image: www.ingimage.com

This book is a translation from the original published under ISBN 978-3-330-76166-7.

Publisher:
Sciencia Scripts
is a trademark of
Dodo Books Indian Ocean Ltd. and OmniScriptum S.R.L publishing group

120 High Road, East Finchley, London, N2 9ED, United Kingdom
Str. Armeneasca 28/1, office 1, Chisinau MD-2012, Republic of Moldova, Europe
Managing Directors: Ieva Konstantinova, Victoria Ursu
info@omniscriptum.com

Printed at: see last page
ISBN: 978-620-8-37518-8

SUMMARY

Inadequate environmental sanitation leads to the emergence of water-borne diseases and environmental degradation. The aim of this study was to obtain information on environmental sanitation in the Sucesso neighbourhood in the municipality of Campo Formoso-BA. During the experimental period, visits were made to the Sucesso neighbourhood, beginning in June 2012 and ending in May 2013. In order to meet the objectives, three sets of actions were carried out: 1) exploratory visits to the neighbourhood to identify the residence where the prototype could be installed; 2) educational interventions to identify the sources of pollution, the distances between pits, monitoring the quality of the water in the cacimbas and alerting residents to suspected contamination; and 3) distribution of information leaflets with a possible solution to the problem, including aspects such as: operation, history and importance of the constructed wetland system (SAC). The results indicated that sanitation was inadequate due to the use of rudimentary pits and the discharge of untreated effluent into water bodies and the use of water from cisterns for human supply. It was recommended to the local population that rudimentary pits should be waterproofed by making them septic tanks, that the distance between septic tanks should be 15 metres and that water from cisterns should be boiled before multiple uses in the home. Visual indicators showed that the constructed wetland system removed suspended and decantable solids, turbidity and odour from the treated domestic sewage. However, the local population pointed to the cost of installation (R$400.00) and operating the system as the main difficulties in using this technology locally. Even with the intense work of mobilising the local population, only a few families went to visit the finished prototype.

Key words: Environment, waste, preservation.

SUMMARY

CHAPTER 1

INTRODUCTION

Water sources are degraded by various anthropogenic actions that jeopardise and limit the availability of water for human consumption. In Brazil, most of the degradation on the urban perimeter is due to organic pollution generated by the lack of sewage collection and treatment services and by the disorderly occupation of the land (IBGE, 2000). However, there has been progress in the basic sanitation services offered to the population (IBGE, 2010).

This problem of pollution and lack of management of water resources means that the consequences of water scarcity are greater in the Brazilian semi-arid region, which faces many difficulties in terms of the lack of water for maintaining the basic needs of the population, as well as for agricultural activities.

The Brazilian semi-arid region covers an area of 969,589.4 km (around 11% of Brazil's territory) and has 12.3% of the country's population (20,858,264 inhabitants) in 1,133 municipalities, which represents 21 inhabitants per km^2 and 22% of Brazil's municipalities (BRASIL, 2005). Because of its demographics and major water scarcity problems, this region is the target of various projects aimed at improving coexistence with drought.

The state of Bahia, located in the Brazilian semi-arid region, like the other states in this region, faces severe periods of drought and due to the degradation of water sources, the population feels the effects of the drought more intensely (BAHIA, 1997; BAHIA, 2006). The lack of management of water resources is the main culprit behind this situation, given that the semi-arid region has freshwater sources, both underground and in river basins, with great potential. This potential needs to be recognised and managed for the benefit of the communities in these regions.

In the municipality of Campo Formoso, located in the northern region of Bahia, there is no municipal sanitation programme, but rather a CODEVASF sanitation programme. The Sucesso neighbourhood in the municipality of Campo Formoso is crossed by the Sucesso River. In this neighbourhood it is common to use artesian wells without any kind of treatment, which is why only 19.81% receive chlorinated water and the public water network serves only 76.42%, leaving 23.58% using water without any kind of treatment, originating from wells and springs without proper supervision. Approximately 83% of families use rudimentary pits and a further 16.98% dispose of faeces and urine in the open and sewage is discharged without any kind of treatment directly into the river, leaving the environment and the population open to contamination (BRASIL, 2005).

CHAPTER 2

OBJECTIVES

2.1. GENERAL OBJECTIVE

Obtaining information on environmental sanitation in the Sucesso neighbourhood in the municipality of Campo Formoso-BA.

2.2. SPECIFIC OBJECTIVES

- Survey the sources of pollution coming from domestic units and carry out a socio-environmental diagnosis of the neighbourhood;
- Propose a flooded system model for local conditions, demonstrating the entire implementation process to the community through home visits and the distribution of information leaflets.

CHAPTER 3

THEORETICAL FRAMEWORK

3.1. ENVIRONMENTAL PRESERVATION

Law 9.433, of 8 January 1997, establishes the National Water Resources Policy, in which the Brazilian government regulates the management of these resources. This law also aims to decentralise this management and involve communities, as it is necessary for public authorities to participate together with communities so that the proper and rational use of water can be extended to future generations (BRASIL, 1997).

Law 10.431 of 20 December 2006 provides for the Environment and Biodiversity Protection Policy of the State of Bahia, article 2º . It establishes the duty of the public authorities and the community to defend, preserve, conserve and restore the environment. In this sense, the study of water availability in all municipalities has been important for understanding how much natural resources are being lost. Currently, municipalities are finding it impossible to utilise the water resources that drain their areas. The inadequate disposal of waste, solid waste and domestic sewage causes accelerated degradation, generating diseases in the population (BAHIA, 2006).

Law 11.445 of 5 January 2007 establishes the legal framework for national basic sanitation. Article 2º lays down the fundamental principles and item I of the same article states that access to basic sanitation must be universalised. Item V also states that the adoption of methods, techniques and processes must take into account local and regional peculiarities

(BRASIL, 2007).

Point VI of the same article also states that basic sanitation must be articulated with urban and regional development, housing, anti-poverty, environmental protection, health promotion and other relevant social interest policies aimed at improving quality of life (BRASIL, 2007).

Article 3° of Chapter I - IV establishes social control, which is a set of mechanisms and procedures that guarantee society information, technical representation and participation in the policy formulation, planning and evaluation processes related to public sanitation services (BRASIL, 2007).

In turn, there is a Municipal Law No. 8/2004, which establishes the environmental code of the Municipality of Campo Formoso, and establishes the fundamental principles of defence, protection, preservation and guarantees of continued use of this resource. Article 2 of this law clearly states that everyone has the right to an ecologically balanced environment and the obligation to defend and preserve it for present and future generations, just like the current Federal Constitution. It also sets out the optimisation and guarantee of the continued use of natural resources, both qualitatively and quantitatively, as a prerequisite for self-sustaining development and the imposition on polluters and predators of the obligation to recover and/or compensate for damage caused by the use of environmental resources for economic purposes; in Title III of Municipal Law No. 8/2004, in Art. 22, it is stated that the Municipality is responsible for implementing the instruments of the Municipal Environmental Policy, in order to achieve the objectives set out in this Code. Therefore, in terms of the legal framework, there are laws in the three spheres of government that deal with the environment and water resources (CAMPO FORMOSO,

2004).

Currently, environmental concerns have become increasingly consistent, as human beings have tried to exceed the natural limits of consumption. In this sense, we realise the importance of preserving water, an essential element for maintaining life. This factor has caused some discomfort in contemporary society, as there is a constant increase in the world's population, industry, agriculture and also livestock farming.

> A phenomenon that can be described as a worldwide movement of environmental awareness appeared suddenly during two years, 1968 to 1970. It seemed that suddenly everyone was concerned about pollution, natural areas, population growth and the consumption of food and energy, according to the widespread coverage of environmental issues in the popular press (ODUM, 1988, p.2).

Several factors contributed to a series of concerns, some of which were: how can the current economic system be maintained without raw materials? How can life be sustained without good quality water sources for human consumption? And how to feed this growing population?

It is known that the problematic issues of today's world are not limited to these questions, but the environment has suffered from indiscriminate practices and concern about this alone will not be enough; more effective practices are needed to minimise these effects.

Due to the growing degradation of the environment and water resources, stricter legislation has been passed in recent years to preserve the quantity and quality of this resource.

3.2. DEGRADATION OF WATER RESOURCES

Water pollution can be conceptualised as the occurrence of phenomena (addition of substances or forms of energy, modifications to the environment) that directly or indirectly alter the nature of a body of water and thus harm its uses. It is important to emphasise that the damage here refers not only to human beings, but also to aquatic biota, social and economic activities in general, natural resources and historical, cultural and landscape collections (BRASIL, 2006). Water pollution can occur in three ways:

- the introduction of artificial and foreign substances into the environment, such as the dumping of pesticides into rivers or contamination by pathogenic organisms;
- the introduction of natural and foreign substances into the environment, such as the addition of sediment to the waters of a lake, reducing its useful volume; and
- a change in the proportion or characteristics of the constituent elements of the environment itself, such as a decrease in the dissolved oxygen content of river water due to the presence of organic matter.

Rivers and lakes that have not been significantly influenced by human activities and that maintain roughly the same natural characteristics are called pristine. There are currently relatively few examples of pristine aquatic environments on our planet, i.e. those virtually untouched by man (VON SPERLING, 2011).

The origin of pollution can be linked to two primary causes: strong population growth and the expansion of industrial activities. The intense population growth seen in recent decades has led to an increase in the generation of sewage and a greater demand for food, which in turn implies a growing consumption of pesticides and fertilisers, which are major polluting agents. This population explosion occurs mainly in the poorest countries, which do not have adequate sewage and pollution control infrastructure, which leads to the establishment of negative conditions for the use of aquatic environments. The expansion of industrial activities, which is also closely related to population growth, has led to the transformation of raw materials into consumer goods on an ever-increasing scale, generating a large quantity of waste, the final destination of which is often *the* aquatic environment (BRASIL, 2006).

Pollution of a body of water is strongly linked to the use made of it. Thus, for example, the presence of nutrients in water is desirable for aquaculture purposes, but extremely harmful when water is taken from lakes or reservoirs for supply (due to eutrophication). Water rich in algae and plants can be used for irrigation, but is inconvenient for electricity generation (clogging of turbines) or even for navigation or recreation. Corrosive water can be used for recreation, but is unsuitable for industrial supply. An aquatic environment with high densities of pathogens cannot be used for recreation, but it can be used for transport. These examples show that the characterisation of water as polluted or not must be linked to its intended use (VON SPERLING, 2011). The main water polluting agents are:

- biodegradable organic matter (sewage, for example), causing oxygen consumption, fish kills and more;

- suspended solids, causing aesthetic problems, sludge deposits, protection from pathogenic organisms, adsorption of pollutants (these adhere to the surface of the suspended solids);

- nutrients, mainly nitrogen and phosphorus, leading to excessive plant growth, the subsequent decomposition of which will jeopardise the water body's oxygen balance;

- pathogens, leading to the emergence of various waterborne diseases;
- non-biodegradable organic matter (pesticides, detergents), leading to bad odours and toxic conditions; and
- heavy metals, also producing toxicity and harming the development of aquatic life.

Pollution of water bodies can be point source, i.*e.* concentrated in space, such as the discharge of sewage pipes into a river, or diffuse, in which pollutants reach the body of water distributed along its length, as *is* the case with pollution from fertilisers and pesticides used in agricultural cultivation (VON SPERLING, 2011).

It is necessary to promote mechanisms to achieve increasingly eco-efficient sustainable development. Some researchers such as Porter and Van der Linde (1995), in studies carried out on large corporations, show how increasing eco-efficiency can lead to better economic and social performance.

In this context, the use of alternative technologies can contribute to improving the environment, since according to the Ministry of Health, degraded water can become a vehicle for spreading diseases among populations and can also jeopardise the practice of agriculture, industry and environmental balance (BRASIL, 2006). Another concern is the lack of sewage collection in Brazil where, according to data from the Ministry of Cities, only 35 per cent of sewage is treated in some way, the rest being discharged directly into water resources without prior treatment (IBGE, 2000).

According to IBGE (2000), pollution of surface water can jeopardise human supply, as these factors lead to compromised water quality.

> The catchments designed and built to take surface water for the supply system, even if they take care with the quality of the source, are subject to the existence of factors that lead to the quality of the water being compromised, such as: the discharge of sanitary sewage, industrial waste, rubbish dumps, mining activities and the presence of pesticide residues (IBGE, 2000, p. 33).

The degradation of water resources is also associated with indiscriminate land use, organic pollution and erosion processes. Salgado and Magalhães Júnior (2006) showed, using turbidity data, the close relationship between land use and impacts on river environments. According to the Ministry of Health, many municipalities and localities do not have the personnel or laboratories capable of monitoring water quality, from the source to the distribution system, and even have difficulties in complying with the requirements of Ministry of Health Ordinance No. 36/1990 (BRASIL, 2006).

3.3MAIN DISEASES CAUSED BY INADEQUATE SANITATION

Most diseases transmitted to humans are caused by microorganisms, particularly viruses, bacteria, protozoa and helminths (intestinal worms). Among the diseases related to water, those transmitted by ingesting contaminated water stand out, and are therefore called waterborne diseases (BRASIL, 2006). Table 1 summarises the main diseases related to inadequate sanitation.

Group of diseases	Forms of transmission	Main diseases	Forms of prevention
Transmitted by the faeco-oral route (food contaminated by faeces)	The pathogenic organism (disease-causing agent) is ingested	Diarrhoea and dysentery Typhoid and typhoid fever Leptospirosis Amebiasis Infectious hepatitis Ascariasis roundworm)	[3]protect and treat water supplies and avoid using contaminated sources [7]Provide adequate water and promote personal, domestic and food hygiene
Controlled by cleaning with water (associated with insufficient water supply)	Lack of water and poor personal hygiene create favourable conditions for its spread	skin and eye infections, such as trachoma, eye-related typhus and scabies	Provide adequate water and promote personal and domestic hygiene
Associated with water (part of the life cycle of the infectious	0 pathogen penetrates	Schistosomiasis	Avoid contact with infected water Protect water sources Adopt appropriate sewage

agent occurs in an aquatic animal)	through the skin or is ingested		disposal measures Combat the host
Transmitted by water-borne vectors	Diseases are spread by insects that hatch in the water or bite near it	Malaria Yellow fever Dengue Filariasis (elephantiasis)	Fighting transmitting insects Eliminate conditions that could favour breeders Avoid contact with breeders Use personal protective equipment

Table 1 - Diseases related to inadequate sanitation.
Source: Brazil (2006).

The occurrence of this type of disease can be minimised or even avoided by adopting appropriate sanitation practices, such as domestic sewage collection and treatment and water supply treatment. Another group of illnesses is associated with a lack of water and the consequent limitations in personal hygiene. Although these diseases are not transmitted by water, they are related to water supply conditions.

There are also diseases, especially worms, whose occurrence is linked to the water environment insofar as part of the life cycle of the infectious agent takes place in the aquatic environment. Finally, the following diseases are worth mentioning
transmitted by water-related vectors, mainly insects that are born or bite in or near water bodies (BRASIL, 2006).

3.4 CLEAN TECHNOLOGIES FOR DOMESTIC SEWAGE TREATMENT

The technologies used by sanitation companies are unfeasible for rural settlements in semi-arid regions, both because of the high cost of implementation and maintenance and because of the wide dispersion of the population in rural areas. Therefore, there is a need to improve low-cost, easy-to-operate technologies for treating domestic sewage, which can increase income by using the treated effluent to fertirrigate agricultural crops with environmental sustainability.

The use of constructed wetland systems (CCS) arose from natural wetland systems, which are areas flooded or saturated by surface or groundwater when it approaches the surface, where the existing vegetation adapts to this condition (MARQUES, 1999). These ecosystems have the capacity to purify the water present in them through physical, chemical and biological mechanisms (MATOS et al., 2010). For this reason, they can be considered as a form of treatment for wastewater, particularly domestic sewage.

The SAC has proven to be efficient in removing organic load, total suspended solids, phosphorus and faecal coliforms from preliminary or primary domestic sewage effluent. However, studies on the use of this wastewater treatment technology in tropical regions have been limited to evaluating the efficiency and use of the treatment process, and there is little information on the existing construction and operational parameters.

SACs are artificial ecosystems with different technologies, using the basic principles of modifying the water quality of natural flooded systems (SALATI, 2011). This system is easy to operate, has a low implementation cost, good results in the treatment of domestic sewage, and the crops produced can be used as animal fodder, ornamental plants and raw materials for handicrafts.

SACs can have different construction characteristics depending on their internal flow (VALENTIM, 2003):

Surface flow: canals have some kind of support material that provides conditions for plant development where the water will flow at small depths;

Subsurface drainage: uses a support medium in the channel that also serves as a filter, which can be made up of gravel, sand and gravel, where the effluent will be discharged for subsequent percolation. The plants grown in these systems are planted directly on the support material. They do not offer conditions for the development and proliferation of mosquitoes or for people and animals to come into contact with the water.

Vertical flow: intermittent flow filters with plant support material, where the water level is below the support medium, making contact with animals and people impossible. It has great potential for nitrification.

Subsurface flow has the following advantages cited by Cooper (1999): the ability to remove suspended solids (SS) and bacteria due to the filtration capacity, removal of Biochemical Oxygen Demand (BOD) greater than the oxygen transfer capacity of plants or the exchange of gases at the air/water interface, and good denitrification capacity (transformation of nitrate into nitrogen gas). Its disadvantage is the possibility of clogging the filter medium with biofilm and solids.

According to Brasil et al. (2005), SACs are a complex wastewater treatment process since pollutants are removed by physical, chemical and biological means as the sewage flows through the system. He also emphasises that the main mechanisms for removing suspended solids (SS) and Biochemical Oxygen Demand (BOD) are flocculation, sedimentation and filtration.

Prochaska et. al. (2008) obtained removals of 52, 60 and 96% in the concentrations of total phosphorus (P $)_{total5}$ nitrate (NO3') and Chemical Oxygen Demand (COD) in a SAC supplied with domestic sewage, using *phragmites australis* as the extractor plant. Brasil (2005) worked with *Typha sp* with two different hydraulic residence times and four types of SACs in the treatment of domestic effluent from septic tanks, obtaining at the end of the experimental trials an average reduction in the values of Chemical Oxygen Demand (COD), suspended solids (SS), turbidity (TB), total nitrogen (NT_{total}) and total phosphorus (P_{total}) of 87, 91, 88, 33 and 35%, respectively.

The choice of plant species is of fundamental importance to the success of wastewater treatment in SACs. According to Marques (1999), the use of local and not exotic species that are tolerant of remaining in a saturated substrate for most of the year should be encouraged. The plant species' functions include extracting the nutrients contained in the wastewater; transferring oxygen to the substrate; providing support (rhizomes and roots) for the growth of bacterial biofilms, as well as improving the permeability of the substrate and the aesthetics of the environment (MATOS et al., 2009).

CHAPTER 4

METHODOLOGY

4.1 STUDY AREA

This work was carried out in the Sucesso neighbourhood in Campo Formoso-BA, located at geographical coordinates "10° 23' 59" to 10° 38' 50" south latitude and "40° 17' 50" to 40° 30'00" west longitude. This municipality is located in the Brazilian semi-arid region, in the north of the state of Bahia, in the territory of Piemonte Norte do Itapicurú, in the micro-region of Senhor do Bonfim, with a population of 66,616 inhabitants *and* a municipal area of 7,259 km. The caatinga predominates in this municipality, with air temperatures ranging from 20 to 30°C, rainfall of up to 490 mm and an evapotranspiration deficit of over 1400 mm (AULER, 1991).

4.2 TYPE OF RESEARCH

The work carried out was identified with action research, a research process that establishes communication with the social research approach, enabling the exchange of information between researchers and the subjects of the conditions being researched (THIOLLENT, 2005). It should be noted that information was collected from the Primary Care Information System (SIAB) and the Maninho Ferreira Family Health Programme (PSF).

Action research arose from the need to bridge the gap between theory and practice and was characterised by the fact that it sought to intervene in practice in an innovative way throughout the entire development process and not just as a possible consequence at the final stage (ENGEL, 2000).

This type of research is special because it allows us to interfere in reality. This interference takes place through discussions and the development of strategies that allow researchers and the subjects of the situation being investigated to seek transformations, clarifications and to overcome problems by working together (THIOLLENT, 2005).

4.3 INSTRUMENTS USED IN THE RESEARCH

The research began with a literature review and exploratory visits to the neighbourhood to identify sources of pollution and also to identify the house where the prototype was installed.

The data provided by the Municipal Health Department was essential for characterising the neighbourhood (APPENDIX I). We also used data collected from residents through questionnaires applied at the end of the experiment (APPENDIX III). The interviewees answered questions about the research initiative, the technology used, the current sanitation situation in the municipality and answered some questions about the treatment process.

The sizing of the prototype constructed flooded system (SAC) considered an average flow of 0.1 m d^{3-1} of domestic sewage (seven people), an application rate of 400 kg of BOD ha^{-1} d^{-1} and a typical domestic sewage BOD concentration of 300 mg L^{-1} following the recommendations of Matos (2007).

$$As = \frac{Q.C}{Ta}.10 \therefore As = \frac{0{,}1.300}{400}.10 \therefore As = 0{,}75\ m^2$$

Where:

As = surface area of the tank, in m^2 ;

Q = inflow, in $m^3 d^{-1}$;

Ta = organic loading rate, kg $ha^{-1} d^{-1}$; and

C = affluent BOD concentration, mg $L .^{-1}$

The SAC was installed between 30th February and 3rd March 2013 in the Sucesso neighbourhood in Campo Formoso-BA. The SAC was built and installed with dimensions of 0.50 metres wide by 1.50 metres long and 0.60 metres deep, with gravel up to 0.40 metres deep. It was waterproofed with polyvinyl chloride geomembrane cut to the dimensions of 1.50 m wide by 2.5 m long and 1.0 m deep.

The effluent from the[1] drum was piped through a PVC pipe with a nominal diameter of 75 mm to a grease trap measuring 250 x 150 x 75 mm. Another pipe with a diameter of 40 mm was derived from this pipe to supply the SAC. A collection box measuring 0.50 m on a side and 0.60 m deep was installed at the outlet of the system with two segments of 40 mm PVC pipe, the first superficial and the second subsurface, to facilitate the collection and removal of the sludge every five days and avoid odours. The SAC was fed with domestic effluent at a flow rate of 12.5 L h^{-1} (100 L applied over an eight-hour period each day). The extractor plant was the aquatic macrophyte *{Eichornia crassipe)* popularly known as baronesa. Figure 1 shows a detail of the container used to store the untreated domestic effluent generated in the household. Figure 2 shows a schematic diagram of the experimental domestic sewage treatment plant. Figure 3 shows a detail of the SAC with the aquatic macrophyte *{Eichornia crassipe)* cultivated in the gravel.

[1] Plastic container with a volume of 200 litres.

Figura 1 - Pump used in the experiment. Source: Researcher's archive (2013).

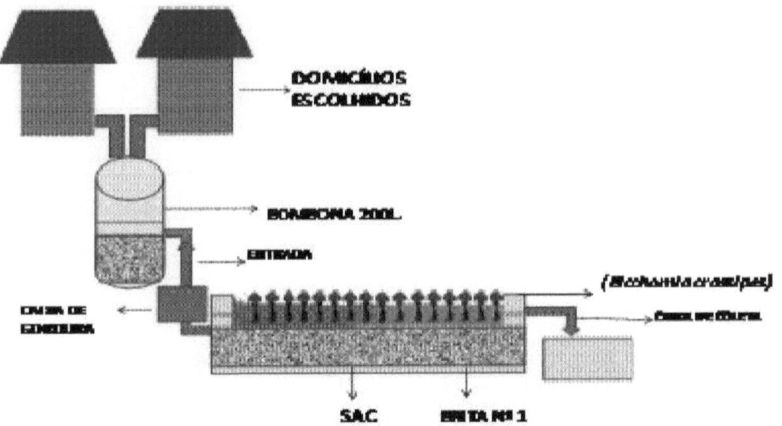

Figura 2 - Schematic diagram of the experimental domestic sewage treatment plant. Source: Researcher's archive (2013).

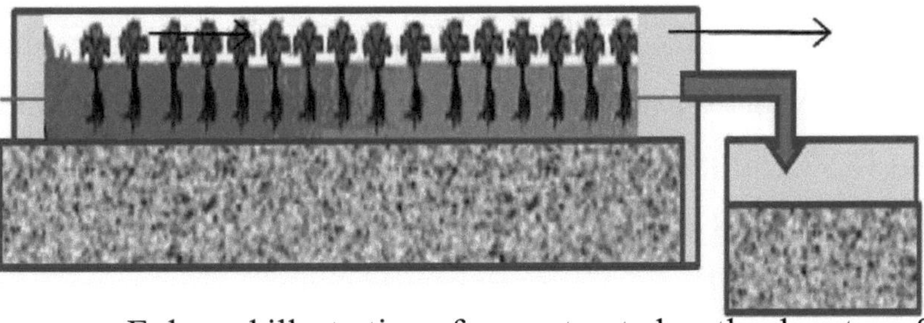

Figura 3 - Enlarged illustration of a constructed wetland system. Source: Researcher's archive (2013).

CHAPTER 5

RESULTS AND DISCUSSIONS

5.1. GENERAL CHARACTERISATION OF THE PUBLIC INVESTIGATED

According to the information collected from the Primary Care Information System (SIAB) and the Maninho Ferreira Family Health Programme (PSF), which serves the entire micro-area of Bairro Sucesso, there are 106 registered families in the Sucesso neighbourhood, with a total of 350 people, 171 male and 179 female. Of the total number of people investigated, 92.16% of those aged between 7 and 14 are in school and 90.76% of those aged 15 or over have completed primary school. This neighbourhood is located in the seat of the municipality of Campo Formoso-BA, but has some rural aspects, as some families have large backyards where they keep chickens, sheep, grow vegetables and so on. The neighbourhood has little physical infrastructure and the predominant professions are bricklayers, self-employed, farmers, maids, seamstresses and others.

The water supply in the neighbourhood is 76.42% from the public network, 23.58% do not have treated water and consume water directly from wells, around 16.98% of these families dispose of faeces/urine in the open, as there is no sewage system, and only 83.02% of these families have rudimentary cesspits. Of this total, only 19.81% of families consume treated water. Annex I contains this information, which was provided by the Campo Formoso-BA Municipal Health Department.

Field visits revealed that residents, because they have few resources and a lack of alternatives, throw their domestic sewage into the River Sucesso without any kind of treatment. This practice is not restricted to the neighbourhood, but due to the geographical location of the river, much of the city's sewage is also disposed of improperly. This jeopardises the health of the city's residents and reduces the quality of the water in the river.

5.2 THE VARIOUS SOURCES OF POLLUTION OF THE RIVER SUCESSO

The visits revealed various sources of pollution such as rudimentary pits and irregular constructions (Figure 4a), domestic sewage discharged into the streets (Figures 4c, 4e and 4f), potteries that contribute to silting up the water body (Figure 4b) and solid household waste (Figure 4e). Figure 4c shows the disposal of domestic sewage in the Sucesso neighbourhood, where the pipes laid in front of the houses discharge the sewage onto the banks of the streets, which then leads to the water body.

It was found that many households use rudimentary pits, but this practice can create a major problem for the community, because due to the large number of springs and the depth of the pits, it was found that the water table is high and is already contaminated by domestic effluent. The distance between the cacimbas and the cesspits was measured and the averages found were between 10 m and 20 m, that is, in 60% of the houses visited, a distance that could lead to contamination.

In the neighbourhood there are 25 cacimbas (groundwater supply wells) and a total of 88 registered cesspits, where most of the rudimentary cesspits present a high risk of contamination, because residents have turned some of these household cacimbas into rudimentary cesspits due to a lack of technical and sanitary knowledge, and this practice can lead to the groundwater of other residents in the neighbourhood.

The rudimentary pits have no waterproofing and no outlet to the complementary treatment unit, so they should be drained periodically, but they have never undergone a maintenance process. They were built without a base lining, leaving water free to drain. The rudimentary pits do not have a ventilation duct to vent the gases, as shown in Figure 4a.

All these problems, which are common in many peripheral areas of Brazilian cities, are a daily reality in the Sucesso neighbourhood, with the aggravating factor that the area has a shallow aquifer, given the existence of springs and wells with water less than 2m from the ground surface.

(a) (b) (c) (d) (e) (f)

Figure 4 - Images of inadequate sanitation in the Sucesso neighbourhood in Campo Formoso-BA, highlighting cesspits without ventilation ducts (a), pottery on the banks of the Sucesso River (b), domestic sewage disposal in the street (c), freshwater spring (d), dumping of solid domestic waste in the Sucesso River (e) and open-air domestic sewage (f).
Source: Researcher's archive (2013).

5.3 ENVIRONMENTAL EDUCATION AND A CONSTRUCTED WETLAND SYSTEM AS AN ALTERNATIVE SOLUTION TO THE PROBLEM OF DOMESTIC SEWAGE IN THE

NEIGHBOURHOOD.

The aim of building the constructed wetland system (SAC) in this work was not to verify the system's efficiency, but to serve as an educational and informational unit for the population of the Sucesso neighbourhood; in addition, an information leaflet (APPENDIX II) was produced to provide information about the system and, from there, to discuss the themes of clean technologies, basic sanitation and public policies. The visits were carried out from the start of the work on 6 June 2012 until the final moments of completion in May 2013. As the neighbourhood has no associative organisation, home visits were chosen.

In the first two visits, the topics of conversation were the sources of pollution identified and raising awareness of the excess rubbish dumped indiscriminately, as shown in Figure 3e.

The next four visits were aimed at explaining to the community the importance of the Sucesso River and the connections that exist within the Itapicuru basin. In this way it was possible to explain that the Sucesso River is perennial, a sub-tributary of the Itapicurú River. As a result, its waters can even contaminate public supply dams. The population was made aware that the situation in the neighbourhood is a national problem, which exists in all Brazilian states and municipalities; however, it is possible to adopt some local measures to minimise the effects of degradation on human health and the environment. To this end, the project involved discussion, reflection and action.

This is not to say that the population of the Sucesso neighbourhood is responsible for the lack of local environmental sanitation. The idea is for people to learn about these techniques, which have been proven to be efficient in scientific circles, and to reflect on the possibilities of these practices being adopted by the government as public policies to improve water resource management and as a form of social inclusion. Law 9.433 of 8 January 1997, which establishes the National Water Resources Policy, aims to decentralise this management and involve communities. Furthermore, the Federal Sanitation Law (BRASIL, 2007) establishes the guidelines for national sanitation. Article I° states that the adoption of methods, techniques and processes must take into account local and regional peculiarities.

In the last three meetings, the aim was to get the neighbourhood's population to follow the process of setting up the SAC, detailing the dimensions used in the system, the half-support and the waterproofing of the system. The process of dehydrating plant biomass to produce animal feed was suggested as a way of utilising the waste generated by the system. In addition, the sludge generated in the pump and the grease trap can be composted, allowing the generation of fertiliser for fruit trees.

During the trial period, the results obtained from the socio-environmental diagnosis were presented and discussed with the population of the Sucesso neighbourhood.

5.4 VISUALISATION OF THE RESULTS OBTAINED BEFORE AND AFTER TREATMENT ON THE SAC.

Figure 5 shows the level of treatment obtained in the experiment, showing that this technology was efficient in treating domestic sewage, as the SAC provides secondary and tertiary level treatment. After treatment, the characteristics suspended solids, decantable solids, odour and turbidity were removed. The plants used in the experiment were collected directly from the river with the collaboration of some local residents. They were also instructed to use other extractive plants such as reeds, which are easily found on the banks of the River Sucesso.

Figure 5 - Visualisation of the level of treatment obtained in the SAC. a) Untreated effluent and b) Treated effluent.
Source: Researcher's archive (2013).

5.5 THE NEIGHBOURHOOD RESIDENTS' DISCOURSE

Interviewee 1: "I liked the initiative, I had never heard of this form of home treatment and I was surprised by the results. "Public bodies should take care of people and the place where they live." "Trying to solve the problem, because the residents of the neighbourhood can't afford it." "The SAC is very interesting, but very labour-intensive. It could solve the problem with the help of trained people." "The cesspits are a necessity for the community. The information given was very important, but we have no way of solving the problem." Doubts: Can the sludge removed from the

SAC also be treated? Researcher's reply: We can treat the sludge from the SAC with an anaerobic biodigester.

Interviewee 2: "I found this new method of sewage treatment interesting, I would like to see this work carried out frequently here in the neighbourhood." "I found the initiative interesting, because the problems in the neighbourhood are enormous and no public body is looking at them." "The sanitation project they started in the city didn't even bother to look at the situation of the pits, the springs or get close to the population." "It stopped and I don't know why". "The SAC is good, but it has a lot of adjustments to make and you always have to change the plants and remove the sludge."

Interviewee 3: "I found it very interesting, I even took part in the process of setting up the SAC, but at first I didn't believe it would work." "Public bodies should give more support to this initiative and try to solve the problem of basic sanitation." "The sanitation project they started seemed good, but nobody ever did anything about it." Question: When treating with the SAC, do you remove the plant that has been used and just discard it? Researcher's reply: You can dehydrate the plant biomass that is removed and grind it up to make animal feed.

Interviewee 4: "I thought it was a good idea, even though I don't understand all the processes involved." "I thought the initiative was good, but it would be good for trained people to monitor the treatment." "The sanitation work they started only served to remove the paving from the streets, because they never did anything else in the town."

Interviewee 5: "I liked it when the pits were measured, and also the explanations given about contamination. I thought the sewage treatment initiative was good, but it's difficult." "Public bodies should try to solve this problem and listen to the residents."

Interviewee 6: "I thought it was a good novelty, but I didn't believe it would work. I'd never seen plants used to clean water before." "It's a good initiative, but looking after this system is a bit of work and it's still not enough to solve the problem, only if the town hall supports it."

Interviewee 7: "I liked the visits and the explanations, I thought the system was good." "We need initiatives like this and also support from public organisations."

Interviewee 8: "Very interesting, I enjoyed learning about this form of sewage treatment. I don't think I could have chosen a better place to develop a project like this". "I would have liked to have taken part in the whole implementation process to motivate other residents to adopt this system." "The measurement of the pits and the information provided were important, as I had already noticed the risks of contamination of the cacimbas." "The town hall's sanitation project in partnership with Codevasf made no difference to the population because it wasn't completed." "The SAC is a good initiative, if it is well monitored and the population gets involved, it could minimise the problems."

Interviewee 9: "I thought the initiative was beautiful, this neighbourhood has a lot of water and we need to find a way to preserve it." "The SAC is a bit of work, but the results were surprising." "The sanitation project that was started didn't listen to the population about the problems in the neighbourhood, it didn't assess the situation of the waterholes and cesspits or the environmental issues. It didn't inform the population why it wasn't completed and left a series of potholes in the streets." "The public authorities should pay more attention to this issue, because these sewers cause illness and harm people's lives."

Interviewee 10: "I'm a health worker and I've never seen this form of sewage treatment before. I found it very interesting, but I don't think it's enough to solve the neighbourhood's problem." "Public bodies should draw up an action plan and invest resources in solving the sanitation problem. The whole city has sewage in the street and even the project they started was interrupted due to lack of funds." "The pits and cesspits need to be periodically assessed because there are risks of contamination. The information given and the measurements taken were good initiatives."

CHAPTER 6

CONCLUSION

With regard to sanitation in the Sucesso neighbourhood, it was found that sewage disposal was inadequate due to the use of rudimentary pits and the discharge of untreated effluent into water bodies and the use of water from cisterns for human supply. It was recommended to the local population that rudimentary pits should be waterproofed by making them septic tanks, that the distance between septic tanks should be 15 metres and that the water from the cisterns should be boiled before multiple uses in the home.

Visual indicators showed that the constructed flooded system removed suspended and decantable solids, turbidity and odour from the treated domestic sewage. However, the local population pointed to the cost of installation (R$400.00) and operating the system as the main difficulties in using this technology locally.

Even with the intense work of mobilising the local population, only a few families went to visit the finished prototype.

CHAPTER 7

REFERENCES

AULER, A. S.; RUBBIOLI, E. L.; MASOTTI, F. S. Methodological evolution in the mapping of Toca da Boa Vista, Campo Formoso/BA. **Espeleo-tema**, Campinas, v. 16, n.l,p.25-39, 1991.

BRAZIL. Ministry of Health. **Impacts on health and the single health system resulting from diseases related to inadequate environmental sanitation.** Brasília: FUNASA, 2010. 248 p. (Final report: studies and research).

BRAZIL. **Law No. 11.445, of 5 January 2007.** Establishes national guidelines for basic sanitation; amends Laws No. 6.766, of 19 December 1979, No. 8.036, of 11 May 1990, No. 8.666, of 21 June 1993, No. 8.987, of 13 February 1995; repeals Law No. 6.528, of 11 May 1978; and makes other provisions. Brasília, 2007. Available at: < http://www.planalto.gov.br/ccivil_03/_ato2007-2010/2007/lei/ll 1445. htm>. Accessed on: 25 May 2011.

BRAZIL. Ministry of Health. Health Surveillance Secretariat. **Surveillance and control of the quality of water for human consumption.** Brasília: Ministry of Health, 2006. 212 p. (Series B. Basic Health Texts).

BRAZIL. Ministry of National Integration. **New delimitation of the Brazilian semi-arid region.** Brasília: 2005, 35p. Available at:< http://www.asabrasil.org.br/UserFiles/File/cartilha_ delimitacao_semi_arido.pdf >. Accessed on: 25 July 2013.

BRASIL, M. S.; MATOS, A. T.; SOARES, A. A.; FERREIRA, P. A. Quality of the effluent from constructed wetland systems used to treat domestic sewage. **Revista Brasileira de Engenharia Agrícola e Ambiental,** Campina Grande, v.9, (Suplemento), p.133-137, 2005.

BRAZIL. **Federal Law No. 9433 of 8 January 1997.** Institutes the national water resources policy, creates the national water resources management system, regulates item XIX of art. 21 of the federal constitution, and amends art. I° of Law no. 8.001, of 13 March 1990, which amended law no. 7.990, of 28 December 1989. Brasília, 1997. Available at:< www.aneel.gov.br/cedoc/bleil9979433.pdf>. Accessed on: 25 May 2013.

BAHIA. Law No. 10.431 of 20 December 2006. Provides for the environment and biodiversity protection policy of the state of Bahia and makes other provisions. Available at: <http://www.jusbrasil.com.br/legislacao/85743/lei-10431-06-bahia- ba>. Accessed on: 25 July 2013.

BAHIA. Secretariat of Water Resources, Sanitation and Housing. **Water resources management plan for the state of Bahia PGRH/BA:** implementation plan. Salvador: 1997. Available at: <http://cbhsaofrancisco.org.br>. Accessed on: 03 March 2013.

CAMPO FORMOSO. **Law No. 18 of June 2004.** Establishes the environmental code for the municipality of Campo Formoso and makes other provisions. Office of the Mayor of Campo Formoso, State of Bahia, p. 1-59. 2004.

COOPER, P. A review of the design and performance of vertical-flow and hybrid reed bed treatment systems. **Water Science and Technology,** London, v.40, n. 3, p. 1-9. 1999.

DANIEL, L. A. **Disinfection processes and alternative disinfectants in the production of drinking water.** Rio de Janeiro: ABES. 2001. 155p.

ENGEL, G. I. **Research-action.** Educar, Curitiba, n. 16, p. 181-191. 2000.

BRAZILIAN INSTITUTE OF GEOGRAPHY AND STATISTICS - IBGE. **National basic sanitation survey 2008.** Rio de Janeiro, 2010. 219p.

BRAZILIAN INSTITUTE OF GEOGRAPHY AND STATISTICS - IBGE. **National basic sanitation survey 2008.** Rio de Janeiro, 2000. 397p.

MARQUES, D. M. Subsurface flow constructed wetlands. In: CAMPOS, R. (Ed.). **Sanitary sewage treatment by anaerobic processes and controlled disposal in soil.** Rio de Janeiro: ABES, 1999. p. 409- 435.

MATOS, A. T.; FREITAS, W. S.; LO MONACO, P. A. V. Efficiency of constructed flooded systems in the removal of pollutants from pig wastewater. **Amb-Água**, Taubaté, v. 5, n 2, p.l 19 -132, 2010.

MATOS, A. T.; ABRAHÃO, S. S.; Lo MONACO; P. A. V.; SARMENTO; A. P.; MATOS; M. P. Extractive capacity of plants in flooded systems used to treat dairy wastewater. **Brazilian Journal of Agricultural and Environmental Engineering.** Campina Grande, v. 14, n. 12, p. 1311-1317, 2009.

MATOS, A. T. **Disposal of wastewater in the soil.** Viçosa, MG: AEAGRI, 2007. 140 p. (Caderno didático n. 38).

ODUM, E. P. **Ecology.** Rio de Janeiro: Guanabara Koogan, 1988, 434p.

PORTER, M. E.; van der LINDE, C. Green and Competitive: ending the stalemate.
Harvard Business Review, p.l 19-134, 1995

PROCHASKA, C. A.; ZOUBOUSLIS, A. I. Treatment performance variation at different depths within vertical subsurface-flow experimental wetlands fed with simulated domestic sewage. **Desalination,** Thessaloniki, v. 237, p.367-377, 2008.

SALATI, E. **Water quality control through constructed wetland systems.** Brazilian Foundation for Sustainable Development. Available a

t<
http://www.fbds.org.br/fbds/Apresentacoes/Controle_Qualid_Agua_Wetlands_ES_ou tOó.pdÊ*. Accessed on 12 March 2011.

SALGADO, A. A. R.; MAGALHÃES JÚNIOR, A. P. Impacts of eucalyptus forestry on the increase in turbidity rates of river water: the case of public supply sources in Caeté-MG. **Geographies,** Belo Horizonte, v. 2, n. 1, p. 47-57, 2006.

SILVA, S. C. **Vertical flow constructed *wetlands* with modified natural soil support in the treatment of domestic sewage.** Brasília: UNB, 2007, 205 f. Thesis (Doctorate in Environmental Technology and Water Resources).

University of Brasília, Brasília.

THIOLLENT, M. **Metodologia da pesquisa-action.** 14 Ed. São Paulo: Cortez, 2005. p.132.

VALENTIM, M. A. A. **Performance of Constructed Wetland beds for sewage treatment:** contributions to design and operation. Campinas-SP: UNICAMP, 2003. 210p.

von S PERL IN G, M. **Introduction to water quality and sewage treatment.** 3.ed. Belo Horizonte: UFMG, 2011. 452p. (Principles of biological wastewater treatment, 1).

ANNEX I

INFORMATION FROM THE MUNICIPAL HEALTH DEPARTMENT 2012

APPENDIX I

EXECUTION SCHEDULE

Target	Home	End	Expected results
• Carry out field visits • Install one (1) SAC prototype on a residence near the river.	06/2012	07/2012	Define all the details of the structural part of the system so that it is as good as possible. operation. Present the treatment with floating aquatic macrophyte *(Eicchorniacrassipes),* such as alternative technology for solving the problem of domestic sewage community.
• Identify sources of pollution and then install the SAC. • Drawing	08/2012	11/2012	Inform the community through information leaflets and home visits about the sources of contamination in the river. Explain how the *wetland* system works and how it can contribute to improving water quality. Provide clarification on the

prototype			methodology used to achieve the project's objectives.
- Elaborate forward to publication01 (one) article scientific.		05/2013	The scientific article will be sent to journals and events in the area in question. In order to socialise the experiences gained during the implementation of the project.
- Culmination of the experiment	05/2013	Up to a month 25/05/2013	Visit to the experiments - Exhibition to families in the local community.

APPENDIX II

INFORMATION LEAFLET

What are *constructed wetlands*?

These are technologically advanced, low-cost systems (Bird, 2004) that can be combined with individual systems to improve the quality of effluent to be discharged into receiving water bodies or reused to irrigate gardens, fields and crops (Silva, 2007).

Constructed wetlands are man-made systems that attempt to mimic the ecological processes found in natural ecosystems (wetlands, floodplains, marshes, swamps or marshes), contributing a variety of biological, social and economic benefits (Ji and Mitchell, 1995).

Constructed wetlands can form a very complete and efficient treatment system (Silvestre and Pedro-de-Jesus, 2002). Their success has already been proven by their history, as the first studies on these systems began in Europe in the 1950s and in the United States in the 1960s (EPA, 1999), with research increasing between 1970 and 1980.

In Brazil, the use of these systems has been spreading slowly and has been studied by various research institutions: the Institute of Applied Ecology (IEA) in Piracicaba, São Paulo; the Agricultural Research and Rural Extension Company of Santa Catarina (EPAGRI); the Agricultural Research Company of Pernambuco (IPA) and public universities - through the Basic Sanitation Research Programme (PROSAB).

Advantages of these systems:

- Low cost;
- High efficiency in improving characteristics physical chemistry of water.
- High biomass production that can be used to produce animal feed (Salati, 2006).

Classification of *constructed* wetland systems

- **Floating** (*Eichhorniacrassipes*),
- **SubmergedElodeacanadensis**, *Elodeanutalli* (Elodea)
- **Emergent** (*Juncusspp and Junco ingens*)

Medium support

Generally, the substrates for these systems are coarse sand, gravel, pebbles and gravel.
The support medium has a dual function: filtration during the wastewater treatment process and support for plant growth.
The physical characteristics (uniformity, porosity and hydraulic conductivity) influence the processes that occur during sewage purification and consequently the performance of the system, so the support medium must be carefully selected (Silva, 2007).

Macrophytes:

Plants play a fundamental role in the treatment, as they provide a surface for microbial films (which carry out most of the treatment) to attach to; they help with filtration and adsorption of wastewater constituents; they transfer oxygen to the water column through their roots and rhizomes and provide thermal insulation (the biomass at the top of the bed helps prevent heat loss through convection (Silvestre and Pedro-de-Jesus, 2002).
To summarise, the form of purification that occurs in systems that use floating plants is due to:

- Adsorption of particles by the root system.
- Absorption of nutrients and metals by the plant.
- By the action of microorganisms associated with the rhizosphere.
- By transporting oxygen to the rhizosphere (Salati, 2006).

Note: All implementation procedures must be supervised by qualified people so that the techniques are properly applied and the results are satisfactory.

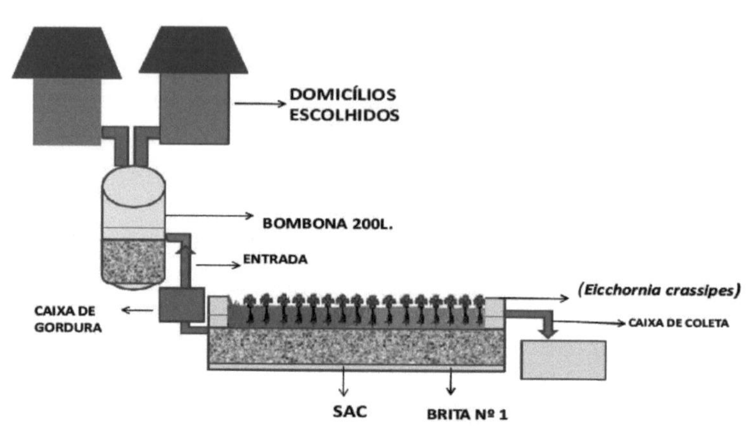

Bibliographical references

SILVA, SELMA CRISTINA **Vertical Flow "*Constructed Wetlands*" with Modified Natural Soil Support in Domestic Sewage Treatment.** WETLAND. Wetlands systems. 2000. Available: http://wwAv.wetlaiid.com.br/.

SALATI, E. Water quality control through a constructed wetlands system. *Brazilian Foundation for Sustainable Development.* Rio de Janeiro Oct. 2006

Silvestre, A. and Pedro-de-Jesus, M. (2002). *Domestic Wastewater Treatment in Artificial Wetlands.*
Final Course Monograph, Instituto Superior Técnico, Department of Biological and Chemical Engineering.

APPENDIX III

QUESTIONNAIRE APPLIED TO RESIDENTS OF THE SUCESSO NEIGHBOURHOOD. Iº A) PROFILE OF THE INTERVIEWEES

Training:

( ) Primary school ( ) Secondary school ( ) Higher education

B) **TYPE OF RESIDENCE ( ) Own home ( ) Rented home ( ) Apartment**

C) **N° OF PEOPLE PER RESIDENCE ()2a4 ()5a8 ()9al0**

( ) Other

D) **Profession**

2º THE SPEECH OF THE HOUSEHOLDERS, a) What do you think of this sewage treatment technology?

b) **What do you think of this initiative in the Neighbourhood?**

c) **What do you think of the way public bodies treat basic sanitation?**

d) **What do you think of the sanitation project initiated by Codevasf in the municipality?**

e) **Do you think the SAC would be enough to solve the neighbourhood's health problem?**

f) **What did you think of the information given about pits and the whole measurement process?**

Printed by Books on Demand GmbH, Norderstedt / Germany